OXFORD
CHINESE
FOR BUSINESS

A Dictionary of Business Terms

Melbourne

OXFORD UNIVERSITY PRESS

Oxford   Auckland   New York

OXFORD UNIVERSITY PRESS AUSTRALIA

Oxford   New York
Athens   Auckland   Bangkok   Bombay
Calcutta   Cape Town   Dar es Salaam   Delhi
Florence   Hong Kong   Istanbul   Karachi
Kuala Lumpur   Madras   Madrid   Melbourne
Mexico City   Nairobi   Paris   Port Moresby
Singapore   Taipei   Tokyo   Toronto
and associated companies in
Berlin   Ibadan

OXFORD is a trade mark of Oxford University Press

Text originally published 1996 as
A Glossary of Common Economic and Business Terms
(English/Chinese) by
Oxford University Press, Singapore

National Library of Australia
Cataloguing-in-publication data:

Chinese for business: a glossary of common
economic and business terms.

ISBN 0 19 550600 6

1. Chinese language – Business Chinese –
Glossaries, vocabularies, etc.

495.181

Printed by Mentor Printers Pte Ltd, Singapore
Published by Oxford University Press
253 Normanby Road, South Melbourne, Australia

Preface

This pocket-sized, user-friendly glossary serves as a quick reference to the common economic and business terms used in English and in Chinese.

This is a useful tool for people who need to carry out business transactions in China and in other places where Chinese is predominantly used.

Comprehensive in its coverage, this glossary includes more than 1100 common economic and business terms such as contracts, currency, credit, interest rate, finance, marketing, trade transaction and insurance.

With China continuing to open up its economy, a working knowledge of the Chinese language and Chinese culture is the key to the vast Chinese market. Readers will find this glossary a useful and handy tool.

序

　　《经贸常用词汇》是一本具崭新内容而又切合实用的工具书。它收集了与经济及贸易有关的1100多条常用词，其内容包括：合同、货币、信贷、利率、金融、销售、交易及保险等。此外，书末的"附录"收集了各国的货币名称。

　　本手册的所有中文词语均附汉语拼音，方便读者拼读。如果一个英文词语有多种不同的释义，则用分号（；）表示区别。

　　对于常到中国或其他使用中文的国家去做生意的人，本手册是极有实用价值的参考书。

Contents
目　录

A Brief Introduction to Hanyu Pinyin

Hanyu Pinyin, known as Pinyin, is a Romanized Chinese alphabet system based on the 26 Roman characters exclusive of **v**. This system was designed by Chinese linguists and adopted by the government of the People's Republic of China (PRC) in 1958, to transcribe the speech sounds of Mandarin and to annotate Chinese characters alphabetically.

The basic unit of a Mandarin sound is called a syllable. The sound of a Chinese character is referred to as a syllable, too. The three components of a Mandarin syllable are:

- the initial consonant
- the final (vowel or nasal vowel)
- tone.

In Mandarin, there are 23 initial consonants inclusive of two semi-vowels, **w** and **y** which are treated as consonants. They are arranged as follows according to position and method of articulation.

1. The initial consonants

1.1 Labials

b	as **p** in	s**p**ort
p	as **p** in	**p**ort
m	as **m** in	**m**an
f	as **f** in	**f**or

1.2 Dentals

d	as **t** in	s**t**ick
t	as **t** in	**t**ear
n	as **n** in	**n**et
l	as **l** in	**l**et

1.3 Gutturals

g	as **k** in	s**k**irt
k	as **k** in	**k**ick
h	as **h** in	**h**er

1.4 Palatals

j	as **j** in	**j**eep
q	as **ch** in	**ch**eer
x	as **x** in	ta**x**i

1.5 Dental sibilants

z	as **z** in	**z**oo
c	as **c** in	**c**ello
s	as **s** in	**s**it

1.6 Retroflexes

zh	as **j** in	**j**ug
ch	as **ch** in	**ch**eer
sh	as **sh** in	**sh**ip
r	as **s** in	lei**s**ure

1.7 Semi-vowels

| w | as **w** in | **w**alk |
| y | as **y** in | **y**es |

2. The final

There are six simple vowels and two consonants
(**-n, -ng**) which form the final in Mandarin.

2.1 Simple vowels

a	o	e	i	u	ü

2.2 Compound vowels

ai	ei	ao	ou	ia	ie	
ua	uo	ui	iu	üe	iao	uai

2.3 Nasal vowels

an	en	in	un	ün		
ang	eng	ing	ong	iong	ian	iang
uan	uang	üan				

2.4 Independent vowel

er

Nine of the above finals are similar to English sounds, e.g:

a	as in	b**a**th
an	as in	pl**an**t
i	as in	k**i**lo
in	as in	**in**
ing	as in	s**ing**
o	as in	**o**val
ong	as in	l**ong**
u	as in	tr**u**th
ua	as in	q**ua**si

The rest of the finals are different from English sounds, e.g:

ü	no corresponding vowel in English, but it is the same as the German **ü**
ai	as in	**eye**
ang	as in	l**ung**

ao	as in	loud
e	as in	her
ei	as in	say
en	as in	burn
eng	as in	her+ng
ou	as in	show
ian	as in	it+end
ie	as in	yes
iu	as in	you
uai	as in	wide
uan	as in	won
uang	as in	woo+lung
ui	as in	way
un	as in	wound
uo	as in	war

3. The tones

Chinese is a tonal language. Basically a syllable is accompanied by a tone which is the variation of pitch. There are four tones in Mandarin, and they are represented by means of tone marks.

the first tone as — in mā (mother)

the second tone as ╱ in má (hemp)

the third tone as ∨ in mǎ (horse)

the fourth tone as ╲ in mà (scold)

Besides the four tones, there is a neutral tone which is so light and short that it does not belong to any of the four tones. The neutral tone does not bear any tone mark.

A

abacus 算盘 suànpan

abbreviation 缩写 suōxiě

abuse of trust 滥用信用 lànyòng xìnyòng，
不顾信用 bùgù xìnyòng

acceptance of contract 承兑合同
chéngduì hétong

account [A/C, a/c] 帐户 zhànghù，户头 hùtóu；
帐目 zhàngmù

accountant 会计师 kuàijìshī

accounting system 会计制度 kuàijì zhìdù

accrued profit 应计利润 yìngjì lìrùn

accumulated profit 累积利润 lěijī lìrùn

across-the-board   全面 quánmiàn

active market   活跃的市场 huóyuède shìchǎng

actual cost   实际成本 shíjì chéngběn

actual value   实际价值 shíjì jiàzhí

administrative expenses
            行政费用 xíngzhèng fèiyòng，
            管理费用 guǎnlǐ fèiyòng

advance   预付 yùfù

advertisement   广告 guǎnggào

advertising agent   广告商 guǎnggàoshāng

advertising expenses   广告费 guǎnggàofèi

advertising media   广告媒介 guǎnggào méijiè

2

advice 通知 tōngzhī

affiliated company 附属公司 fùshǔ gōngsī

agency agreement 代理协定 dàilǐ xiédìng

agency contract 代理合同 dàilǐ hétong

air freight 航空运费 hángkōng yùnfèi；
航空货运 hángkōng huòyùn

air risks 空运险 kōngyùnxiǎn

airmail 航空邮件 hángkōng yóujiàn

airmail postage 航空邮费 hángkōng yóufèi

all risks 一切险 yīqièxiǎn，
综合险 zōnghéxiǎn

allotment 分配 fēnpèi

allowance 津贴 jīntiē

allowance and rebates 折让和回扣
zhéràng hé huíkòu

amendment 修改 xiūgǎi

alternative duty 选择税 xuǎnzéshuì

amount 金额 jīn'é；总数 zǒngshù

amount in figures 小写金额 xiǎoxiě jīn'é

amount in words 大写金额 dàxiě jīn'é

analysis 分析 fēnxī

announcement 公布 gōngbù；通告 tōnggào

annual income 年收入 niánshōurù

annual interest 年息 niánxī，年利息 niánlìxī

annual report　年报 niánbào，
常年报告书 chángnián bàogàoshū

annuity　年金 niánjīn

application　申请 shēnqǐng；
申请书 shēnqǐngshū

application form　申请表 shēnqǐngbiǎo，
申请表格 shēnqǐng biǎogé

appoint　指定 zhǐdìng；委任 wěirèn

appraisal　估价 gūjià；评定 píngdìng

appropriation　拨款 bōkuǎn；挪用 nuóyòng

approval　批准 pīzhǔn，核准 hézhǔn

arbitrage　套汇 tàohuì，套利 tàolì

arbitration　仲裁 zhòngcái

arrears　拖欠 tuōqiàn，余额 yú'é

article　商品 shāngpǐn；条款 tiáokuǎn

Asian Currency Unit　亚元单位
　　　　　　　　　　Yàyuán Dānwèi

assess　审估 shěngū，估值 gūzhí

assets　资产 zīchǎn，财产 cáichǎn

assurance　保险 bǎoxiǎn；担保 dānbǎo

attachment　附件 fùjiàn

auction　拍卖 pāimài

audit　审帐 shěnzhàng，审计 shěnjì

auditor　审计师 shěnjìshī，
　　　　　查帐员 cházhàngyuán

authority to purchase (A/P)
委托购买证 wěituō gòumǎizhèng

automated teller machine (ATM)
自动提款机 zìdòng tíkuǎnjī，
自动取款机 zìdòng qǔkuǎnjī

available　可利用的 kělìyòngde；
生效的 shēngxiàode

average cost　平均成本 píngjūn chéngběn

average rate　平均收费率 píngjūn shōufèilǜ

award　授予 shòuyǔ

B

back order   拖欠定货 tuōqiàn dìnghuò，
              延交定货 yánjiāo dìnghuò

back pay   欠薪 qiànxīn，
              拖欠工资 tuōqiàn gōngzī

backlog   积压未交付的订货
              jìyā wèi jiāofù de dìnghuò

bad debts   烂帐 lànzhàng，呆帐 dāizhàng

balance   余额 yú'é；差额 chā'é；
              平衡 pínghéng

balance sheet [B.S., b.s.]
              资产负债表 zīchǎn fùzhàibiǎo

bank   银行 yínháng

bank credit   银行信贷 yínháng xìndài

bank draft [B/D] 银行汇票 yínháng huìpiào

bank notes 钞票 chāopiào，纸币 zhǐbì

bank rate of discount 银行贴现率
yínháng tiēxiànlǜ

bank rate of exchange 银行汇率 yínháng huìlǜ

bank overdraft 银行透支 yínháng tòuzhī

banker's acceptance 银行承兑
yínháng chéngduì

bankrupt 破产者 pòchǎnzhě；
破产的 pòchǎnde

bankruptcy 破产 pòchǎn，倒闭 dǎobì

bargain 交易 jiāoyì；便宜物 piányiwù；
讨价还价 tǎojià huánjià，
议价 yìjià

bargaining power 讨价还价能力
tǎojià huánjià nénglì

barter trade 易货贸易 yìhuò màoyì

basic price 基价 jījià

batch 批 pī，整批 zhěngpī

bazaar 集市 jíshì，市场 shìchǎng

bear 卖空者 màikōngzhě，空头 kōngtóu

bear market = bearish market 淡市 dànshì，
熊市 xióngshì，跌市 diēshì

bearer 持票人 chípiàorén

beneficiary 受益人 shòuyìrén

benefit 利益 lìyì；津贴 jīntiē

best-seller   畅销品 chàngxiāopǐn；
　　　　　　畅销书 chàngxiāoshū

bid   投标 tóubiāo

bid price   标价 biāojià

bilateral trade   双边贸易 shuāngbiān màoyì

bilateral treaty   双边条约 shuāngbiān tiáoyuē

bill   帐单 zhàngdān；票据 piàojù；
　　　汇票 huìpiào

bill of lading [B/L]   提单 tídān

black market   黑市 hēishì

black money   黑钱 hēiqián

blackmail   勒索 lèsuǒ

blank cheque　空白支票 kòngbái zhīpiào，
空额支票 kòng'é zhīpiào

blanket order　总括定单 zǒngkuò dìngdān

block　封锁 fēngsuǒ；
限制使用 xiànzhì shǐyòng

block deal　大宗股票交易
dàzōng gǔpiào jiāoyì

blockage　封锁状态 fēngsuǒ zhuàngtài

blue chip = blue-chip share　蓝筹股 lánchóugǔ

blue-collar workers　蓝领工人 lánlǐng gōngrén

blueprint　蓝图 lántú

board　委员会 wěiyuánhuì；
董事会 dǒngshìhuì；理事会 lǐshìhuì

board of directors　董事会 dǒngshìhuì

bogus money　伪币 wěibì

bond　公债 gōngzhài，债券 zhàiquàn

bond market　债券市场 zhàiquàn shìchǎng

bonus share　红股 hónggǔ

bonus system　职工奖金制度
　　　　　　zhígōng jiǎngjīn zhìdù

book value　帐面价值 zhàngmiàn jiàzhí

booking　预定 yùdìng；订货 dìnghuò

boom　兴旺 xìngwàng；利市 lìshì；
　　　（股价）高涨 (gùjià) gāozhǎng

borrower　借钱人 jìeqiánrén；
　　　　　借用者 jìeyòngzhě

bottom price　最低价 zuìdījià

boycott　联合抵制 liánhé dǐzhì

branch office　分公司 fēngōngsī，
　　　　　　　　分行 fēnháng，分店 fēndiàn

breach of contract　毁约 huǐyuē，
　　　　　　　　　　违反合同 wéifǎn hétong

break even　不赔不赚 bùpéi bùzhuàn，
　　　　　　　保本 bǎoběn

brisk market　市场活跃 shìchǎng huóyuè

broker　经纪人 jīngjìrén

brokerage　经纪人佣金 jīngjìrén yòngjīn

brought forward [b/f]　承前页 chéng qiányè，
　　　　　　　　　　　　承上页 chéng shàngyè

14

budget year　预算年度 yùsuàn niándù

bulk　散装 sǎnzhuāng

bull　买空者 mǎikōngzhě；多头 duōtóu

bull market　旺市 wàngshì，牛市 niúshì

buoyant　活跃的 huóyuède

business　商业 shāngyè；生意 shēngyì；
公事 gōngshì；业务 yèwù；
营业 yíngyè

business centre　商业中心 shāngyè zhōngxīn

business district　商业区 shāngyèqū

business guide　商业指南 shāngyè zhǐnán

business hours　营业时间 yíngyè shíjiān

business tax 营业税 yíngyèshuì

buyer 买方 mǎifāng，买主 mǎizhǔ，
买家 mǎijiā

buying on margin 抵押购买 dǐyā gòumǎi

by-law 公司章程 gōngsī zhāngchéng；
附则 fùzé

by-product 副产品 fùchǎnpǐn

C

call 催交股款 cuījiāo gǔkuǎn；
购买权 gòumǎiquán；
停靠（港口）tíngkào(gǎngkǒu)；
（一次）电话 (yīcì) diànhuà

call option 购买选择权 gòumǎi xuǎnzéquán

capacity 生产能力 shēngchǎn nénglì；
容量 róngliàng

capital 资本 zīběn

capital invested 投入资本 tòurù zīběn

capital stock 股本 gǔběn

carbon copy [c.c.] 副本 fùběn

cargo 船货 chuánhuò；货物 huòwù

cargo vessel　货船 huòchuán，货轮 huòlún

carriage　货运 huòyùn，运输 yùnshū；
运费 yùnfèi

carried forward　转入次页 zhuǎnrù cìyè

cash　现金 xiànjīn，现款 xiànkuǎn；
兑现 duìxiàn，付现款 fù xiànkuǎn

cash advance　预付现金 yùfù xiànjīn

cash flow　现金流动 xiànjīn liúdòng

cash market　现金交易市场
xiànjīn jiāoyì shìchǎng，
现货市场 xiànhuò shìchǎng

cash on delivery [C.O.D.]
交货付款 jiāohuò fùkuǎn，
现款交货 xiànkuǎn jiāohuò

cash order [C/O]   现金订货 xiànjīn dìnghuò

cashier   出纳员 chūnàyuán，
          收支员 shōuzhīyuán

cashier order   银行本票 yínháng běnpiào

ceiling   最高限额 zuìgāo xiàn'é

ceiling price   最高价格 zuìgāo jiàgé

certificate   证书 zhèngshū；执照 zhízhào

certificate of deposit [C.D.]
          定期存款单 dìngqī cúnkuǎndān

certificate of entitlement [COE]
          拥车证 yōngchēzhèng

Central Provident Fund [CPF]
          中央公积金（公积金）
          zhōngyāng gōngjījīn (gōngjījīn)

chain stores　连锁商店 liánsuǒ shāngdiàn

chairman　主席 zhǔxí；会长 huìzhǎng

Chamber of Commerce [C.C.]　商会 shānghuì

channel of distribution　分销渠道 fēnxiāo qúdào，
　　分销系统 fēnxiāo xìtǒng

chart　图表 túbiǎo

chartered accountant　特许会计师 tèxǔ kuàijìshī

check = cheque　支票 zhīpiào

chop　图章 túzhāng；牌号 páihào；
　　出港证 chūgǎngzhèng

circular　通告 tōnggào，传单 chuándān

circulation　流通 liútōng；流通额 liútōng'é；
　　发行额 fāxíng'é

claim　索赔 suǒpéi

classification　分类 fēnlèi；分级 fēnjí

clearance of goods　报关 bàoguān

clearance sale　清仓大减价 qīngcāng dàjiǎnjià

clerical error　笔误 bǐwù，记录错误 jìlù cuòwù

client　客户 kèhù；委托人 wěituōrén

close an account　结清帐户 jiéqīng zhànghù，
结束帐户 jiéshù zhànghù

closing date　截止日期 jiézhǐ rìqī

closing price　收市价 shōushìjià，
收盘价 shōupánjià

closing rate　收盘汇率 shōupán huìlǜ

co-borrower 共同借款人 gòngtóng jièkuǎnrén

cold store 冷藏仓库 lěngcáng cāngkù

collapse （价格、股市）暴跌
(jiàgé, gǔshì) bàodiē

collateral 抵押品 dǐyāpǐn，抵押 dǐyā

collateral security 附属担保品 fùshǔ dānbǎopǐn

commercial circles 商界 shāngjiè，
商业界人士
shāngyèjiè rénshì

commercial invoice 商业发票 shāngyè fāpiào

commercial services 商业服务 shāngyè fúwù

commission charges 手续费 shǒuxùfèi，
佣金 yòngjīn

commodity 商品 shāngpǐn；
农产品 nóngchǎnpǐn

common tariff 共同关税 gòngtóng guānshuì

Community Chest 公益金 Gōngyìjīn

compensation 补偿 bǔcháng；赔偿 péicháng；
赔偿金 péichángjīn

competition 竞争 jìngzhēng

complete plant 成套设备 chéngtào shèbèi

composite price 综合价格 zōnghé jiàgé

compound interest 复利 fùlì

computer 电脑 diànnǎo，
电子计算机 diànzǐ jìsuànjī

confidential 机密 jīmì

conference　会议 huìyì

confirmation　确认 quèrèn，核实 héshí，
证实 zhèngshí

consignment　寄售 jìshòu，寄销 jìxiāo

consumer goods　消费品 xiāofèipǐn，
消费货物 xiāofèi huòwù

consumer price　消费品价格 xiāofèipǐn jiàgé

consumption　消费 xiāofèi；消耗量 xiāohàoliàng

contra　对销 duìxiāo；对敲 duìqiāo

contract　合同 hétong，契约 qìyuē

contract note　股票买卖通知书
gǔpiào mǎimài tōngzhīshū；
买卖合同 mǎimài hétong

contractor 承包商 chéngbāoshāng

conversion rate 换算率 huànsuànlǜ；
汇兑率 huìduìlǜ，汇价 huìjià

convertible currency 可兑换货币
kěduìhuàn huòbì

copy 副本 fùběn；抄本 chāoběn

copyright 版权 bǎnquán，著作权 zhùzuòquán

corporate law 公司法令 gōngsī fǎlìng

cost 成本 chéngběn

cost and freight [C&F, CF]
成本加运费 chéngběn jiā yùnfèi

cost, insurance and freight [CIF]
到岸价格 dào'àn jiàgé

cost of maintenance　维修费 wéixiūfèi

cost of production　生产成本
shēngchǎn chéngběn

counterfeit notes　伪钞 wěichāo

counter offer　还盘 huánpán，反发价 fǎnfājià

coupon　息票 xīpiào；赠券 zèngquàn；
固本 gùběn

credit　信用 xìnyòng；信贷 xìndài

credit card　信用卡 xìnyòngkǎ

credit limit = credit line　信贷限额 xìndài xiàn'é

credit loan　信用贷款 xìnyòng dàikuǎn

credit restriction　信贷限制 xìndài xiànzhì

credit terms 信贷条件 xìndài tiáojiàn

creditor 债权人 zhàiquánrén，债主 zhàizhǔ

crossed cheque 划线支票 huàxiàn zhīpiào

cum bonus 附有红利 fùyǒu hónglì

cum dividend [c.d.] 附有股息 fùyǒu gǔxī

currency 通货 tōnghuò，货币 huòbì

currency appreciation 货币升值 huòbì shēngzhí

currency depreciation 货币贬值 huòbì biǎnzhí

currency unit 货币单位 huòbì dānwèi

current account 往来帐户 wǎnglái zhànghù，
来往帐户 láiwǎng zhànghù，
活期存款帐户
huóqī cúnkuǎn zhànghù

customer 顾客 gùkè，客户 kèhù

customs duty 关税 guānshuì

customs invoice 海关发票 hǎiguān fāpiào

D

daily necessities　日常必需品 rìcháng bìxūpǐn

damage　损害 sǔnhài，毁坏 huǐhuài，
损失 sǔnshī

dangerous goods　危险货物 wēixiǎn huòwù

date of delivery　交货日期 jiāohuò rìqī

date of expiration = date of expiry
期满日 qīmǎnrì，到期日 dàoqīrì

date terms　日期条件 rìqī tiáojiàn

dateline　最后限期 zuìhòu xiànqī

deal　交易 jiāoyì，买卖 mǎimài

debenture 无担保债券 wúdānbǎo zhàiquàn；
退税凭单 tuìshuì píngdān

dealer 证券商 zhèngquànshāng；
交易员 jiāoyìyuán；
经销人 jīngxiāorén

debit 借入 jièrù；借记 jièjì；
记入借方的款项 jìrù jièfāng de kuǎnxiàng

debt 债务 zhàiwù；欠款 qiànkuǎn

debtor 债务人 zhàiwùrén，负债者 fùzhàizhě

decimal system 十进制 shíjìnzhì，公制 gōngzhì

declaration 申报 shēnbào；声明 shēngmíng；
报单 bàodān

deduction 扣除 kòuchú；扣除额 kòuchú'é

default 违约 wéiyuē；拖欠 tuōqiàn

defer 延期 yánqī，推迟 tuīchí

deficit 亏损 kuīsǔn；赤字 chìzì；逆差 nìchā

deliver 交付 jiāofù；递送 dìsòng

delivery order [D/O] 提货单 tíhuòdān，
交货单 jiāohuòdān

demand draft [D/D] 即期汇票 jíqī huìpiào

denomination 面额 miàn'é

deposit 押金 yājīn；定金 dìngjīn；
存款 cúnkuǎn

depreciation 折旧 zhéjiù

derivatives trading 衍生产品交易
yǎnshēng chǎnpǐn jiāoyì

description 规格 guīgé；品种 pǐnzhǒng

31

devaluation 贬值 biǎnzhí

direct tax 直接税 zhíjiēshuì

discharge 卸货 xièhuò；解除 jiěchú；
解雇 jiěgù

discount 折扣 zhékòu；贴现 tiēxiàn

discount rate 贴现率 tiēxiànlǜ，
贴水率 tiēshuǐlǜ，
折扣率 zhékòulǜ

dishonoured cheque 空头支票
kōngtóu zhīpiào，
拒付支票 jùfù zhīpiào

display 陈列 chénliè

distribution 分配 fēnpèi；销售 xiāoshòu；
分红 fēnhóng

distributor 经销商 jīngxiāoshāng，
经销人 jīngxiāorén

dividend 股息 gǔxī，股利 gǔlì

dock charges 码头费 mǎtoufèi

dockage 码头费 mǎtoufèi，船坞费 chuánwùfèi

document 文件 wénjiàn；证件 zhèngjiàn；
单据 dānjù

documents against acceptance [D/A]
承兑交单 chéngduì jiāodān

documents against payment [D/P]
付款交单 fùkuǎn jiāodān

dollar crisis 美元危机 Měiyuán wēijī

domestic market 国内市场 guónèi shìchǎng

domestic trade　国内贸易 guónèi màoyì，
内贸 nèimào

double taxation　双重课税 shuāngchóng kèshuì

down payment　首期付款 shǒuqī fùkuǎn；
订金 dìngjīn

draft　汇票 huìpiào

drawee　受票人 shòupiàorén，付款人 fùkuǎnrén

drawer　出票人 chūpiàorén，开票人 kāipiàorén

dumping　倾销 qīngxiāo

dutiable goods　征税货物 zhēngshuì huòwù

duty-free goods　免税商品 miǎnshuì shāngpǐn

E

earned income   劳动收入 láodòng shōurù，
          执业所得 zhíyè suǒdé

earning per share [EPS]   每股盈利 měigǔ yínglì

economic development   经济发展 jīngjì fāzhǎn

economic forecasting   经济预测 jīngjì yùcè

economic recession   经济衰退 jīngjì shuāituì

effective date   生效日期 shēngxiào rìqī，
          有效日期 yǒuxiào rìqī

embargo   禁运 jìnyùn；禁止进出口令
          jìnzhǐ jìn-chūkǒu lìng

employee   雇员 gùyuán；职工 zhígōng

employer   雇主 gùzhǔ

en route　在途中 zài túzhōng

endorse　背签 bèiqiān，背书 bèishū

engage　雇用 gùyòng；保证 bǎozhèng；
预定 yùdìng

enterprise　企业 qǐyè

entertainment　招待 zhāodài；娱乐 yúlè

entertainment tax　娱乐税 yúlèshuì

entrepreneur　企业家 qǐyèjiā

entrepôt trade　转口贸易 zhuǎnkǒu màoyì

entry formalities　入境手续 rùjìng shǒuxù；
进口手续 jìnkǒu shǒuxù

equipment　设备 shèbèi，装备 zhuāngbèi，
器材 qìcái

equities 股票 gǔpiào，证券 zhèngquàn

equity investment 股本投资 gǔběn tóuzī

error 差错 chācùo，错误 cuòwù

establish 设立 shèlì；开立 kāilì；开拓 kāituò

estate 产业 chǎnyè，房地产 fángdìchǎn

estimate 估价 gūjià；估计 gūjì

European Economic Community [E.E.C.]
欧洲经济共同体
Ōuzhōu Jīngjì Gòngtóngtǐ

European Monetary Agreement [E.M.A.]
欧洲货币协定 Ōuzhōu Huòbì Xiédìng

exchange 交换 jiāohuàn；交易 jiāoyì；
兑换 duìhuàn；交易所 jiāoyìsuǒ

exchange broker　外汇经纪人 wàihuì jīngjìrén

exchange crisis　外汇危机 wàihuì wēijī

exchange control　外汇管制 wàihuì guǎnzhì

exchange control regulations
　　　　外汇管制条例 wàihuì guǎnzhì tiáolì

exchange fluctuation
　　　　汇价变动 huìjià biàndòng,
　　　　汇率波动 huìlǜ bōdòng

exchange rate　兑换率 duìhuànlǜ, 汇率 huìlǜ,
　　　　汇价 huìjià

exchange risk　外汇风险 wàihuì fēngxiǎn,
　　　　汇兑风险 huìduì fēngxiǎn

exchange tax　外汇税 wàihuìshuì

execute a contract 履行合同 lǚxíng hétong，
执行合约 zhíxíng héyuē

exemption 豁免 huòmiǎn

expenditure 支出 zhīchū；费用 fèiyòng；
开销 kāixiāo

expiration = expiry 期满 qīmǎn，到期 dàoqī

export 出口 chūkǒu，输出 shūchū

export duty 出口税 chūkǒushuì

export procedures 出口程序 chūkǒu chéngxù

export sales 外销 wàixiāo

exposition 博览会 bólǎnhuì

express goods 快运货 kuàiyùnhuò

extension 延期 yánqī，扩大 kuòdà

external trade 对外贸易 duìwài màoyì

F

face value　（票据）面值 (piàojù) miànzhí

facilities　设备 shèbèi；便利 biànlì

fascimile [fax]　传真 chuánzhēn；
传真本 chuánzhēnběn

factory cost　工厂成本 gōngchǎng chéngběn

fair　公平的 gōngpíngde；定期集市
dìngqī jíshì；展览会 zhǎnlǎnhuì；
义卖市场 yìmài shìchǎng

falling market　市价下跌 shìjià xiàdiē；
价格下跌的市场
jiàgé xiàdiē de shìchǎng

fashion goods　时髦商品 shímáo shāngpǐn

favourable balance　顺差 shùnchā

fee 手续费 shǒuxùfèi；会费 huìfèi；
酬金 chóujīn

field sales 现场销售 xiànchǎng xiāoshòu

filing 归档 guīdàng

finance markets 金融市场 jīnróng shìchǎng

financial analysis 财务分析 cáiwù fēnxī

financial investment 金融投资 jīnróng tóuzī

financial policy 财政政策 cáizhèng zhèngcè，
金融政策 jīnróng zhèngcè

financial report 财务报告 cáiwù bàogào，
财政报告 cáizhèng bàogào

financial statement 财务报表 cáiwù bàobiǎo

financial year = fiscal year
　　　　财政年度 cáizhèng niándù，
　　　　会计年度 kuàijì niándù

finder　　中人 zhōngrén

fire insurance　　火灾保险 huǒzāi bǎoxiǎn，
　　　　　　火险 huǒxiǎn

firm　　坚挺的 jiāntǐngde，稳定的 wěndìngde；
　　　　确定的 quèdìngde

fixed assets　　固定资产 gùdìng zīchǎn

fixed capital　　固定资本 gùdìng zīběn

fixed deposit　　定期存款 dìngqī cúnkuǎn

fixed price　　固定价格 gùdìng jiàgé

fixed exchange rate　　固定汇率 gùdìng huìlǜ

floating exchange rate 浮动汇率 fúdòng huìlǜ

floor price 最低限价 zuìdī xiànjià，底价 dǐjià

fluctuating rate 波动汇率 bōdòng huìlǜ

fluctuation in prices 价格波动 jiàgé bōdòng

follow-up 连续广告 liánxù guǎnggào；
继续的 jìxùde

foodstuff 食品 shípǐn

foot 英尺 yīngchǐ，呎 chǐ

forced sale 强制销售 qiángzhì xiāoshòu

forced selling 强制售卖股票
qiángzhì shòumài gǔpiào，
逼仓 bīcāng

forecast 预测 yùcè

foreign currency 外币 wàibì

foreign exchange [F/X] 外汇 wàihuì

foreign exchange budget 外汇预算 wàihuì yùsuàn

foreign exchange control
外汇管制 wàihuì guǎnzhì

foreign exchange quotations
外汇牌价 wàihuì páijià

foreign exchange reserves
外汇储备 wàihuì chǔbèi

foreign trade 对外贸易 duìwài màoyì

forfeit 没收 mòshōu；丧失（权利）
sàngshī (quánlì)；罚金 fájīn；没收物
mòshōuwù

forged document　伪造的文件
wěizào de wénjiàn，
伪造单据 wěizào dānjù

forward exchange　期货外汇 qīhuò wàihuì

forward transaction　期货交易 qīhuò jiāoyì

fractional money　辅币 fǔbì

franchise　特许权 tèxǔquán；
专营权 zhuānyíngquán

fraud　欺诈 qīzhà，欺诈行为 qīzhà xíngwéi

free convertibility　自由兑换性
zìyóu duìhuànxìng

free exchange rate　自由汇率 zìyóu huìlǜ

free from all average [f.a.a.]
全损赔偿 quánsǔn péicháng

free from particular average [FPA]
平安险 píng'ānxiǎn

free market price 自由市场价格
zìyóu shìchǎng jiàgé

free of charge 免费 miǎnfèi

free on board [F.O.B.]
船上交货价 chuánshàng jiāohuòjià，
离岸价格 lí'àn jiàgé

free port 免税港 miǎnshuìgǎng，
自由港 zìyóugǎng

free trade 自由贸易 zìyóu màoyì

freeze 冻结 dòngjié

freight charges 运费 yùnfèi

freight in full 全包运费 quánbāo yùnfèi

freight rate　运费率 yùnfèilǜ

freight tariff　运费率表 yùnfèilǜ biǎo，
运价表 yùnjiàbiǎo

frozen capital　冻结资金 dòngjié zījīn

full-time service　专任职务 zhuānrèn zhíwù

function　职能 zhínéng；功能 gōngnéng

fund　资金 zījīn；基金 jījīn

futures　期货 qīhuò

futures trading　期货交易 qīhuò jiāoyì

G

gain　营利 yínglì，获利 huòlì

gains　利润 lìrùn，收益 shōuyì

gallon　加仑 jiālún

general expenses　日常费用 rìcháng fèiyòng，
一般费用 yībān fèiyòng；
管理费用 guǎnlǐ fèiyòng

general ledger　总分类帐 zong-fēnlèizhàng

general meeting of shareholders
股东大会 gǔdōng dàhuì

general tariff　一般税率 yībān shuìlǜ，
普通税率 pǔtōng shuìlǜ

genuine goods　真货 zhēnhuò

gift coupon　礼品券 lǐpǐnquàn，赠券 zèngquàn

gift shop　礼品店 lǐpǐndiàn

glut　充斥 chōngchì，供过于求 gōng guòyú qiú，
市场滞销 shìcháng zhìxiāo

go down　减低 jiǎndī，下降 xiàjiàng

godown　仓库 cāngkù，货栈 huòzhàn

going rate　现率 xiànlǜ

gold and dollar reserves
黄金及美元储备
huángjīn jí Měiyuán chǔbèi

gold and foreign exchange reserves
黄金及外汇储备 huángjīn jí wàihuì chǔbèi

gold bar　金条 jīntiáo

gold rush　抢购黄金 qiǎnggòu huángjīn

Government-link Company [GLC]
　　　　　政联公司 zhènglián gōngsī

Goods and Service Tax [GST]
　消费税 xiāofèishuì

grace period　宽限期 kuānxiànqī

grading　分级 fēnjí

grant　授予 shòuyǔ；补助金 bùzhùjīn，
　　　　资助金 zīzhùjīn

greenback　美钞 Měichāo

gross earnings　毛利 máolì，总收益 zǒngshōuyì

gross import value　进口总值 jìnkǒu zǒngzhí

gross export value　出口总值 chūkǒu zǒngzhí

gross income   总收入 zǒng shōurù

gross national product [GNP]
   国民生产总值
   guómín shēngchǎn zǒngzhí

gross profit   毛利润 máolìrùn

gross weight   毛重 máozhòng，总重 zǒngzhòng

group insurance   集体保险 jítǐ bǎoxiǎn

growth   增长 zēngzhǎng

guarantee   保证 bǎozhèng；保证书
   bǎozhèngshū；抵押品 dǐyāpǐn

guarantor   保证人 bǎozhèngrén，担保人
   dānbǎorén

H

half price   半价 bànjià

hand-made   手工制作的 shǒugōng zhìzuò de

handbook   手册 shǒucè

handicraft   手工 shǒugōng；手工业 shǒugōngyè

handle   经营 jīngyíng，买卖 mǎimài；
处理 chǔlǐ

HANDLE WITH CARE
小心轻放 xiǎoxīn qīngfàng

harbour dues   入港税 rùgǎngshuì；港务费
gǎngwùfèi

hard cash   现金 xiànjīn

hardware   硬件 yìngjiàn

head office　总公司 zǒnggōngsī，总部 zǒngbù，
总行 zǒngháng

hedge　套期保值 tàoqī bǎozhí，
套头交易 tàotóu jiāoyì，
对冲买卖 duìchōng mǎimài

hire purchase　分期付款购买
fēnqī fùkuǎn gòumǎi

hoard　囤积 túnjī；储藏 chǔcáng

hold in pledge　抵押 dǐyā

holder　持票人 chípiàorén，持有人 chíyǒurén

holding company　控股公司 kònggǔ gōngsī

home-made　国产的 guóchǎnde；自制的 zìzhìde

home market　国内市场 guónèi shìchǎng

horsepower　马力 mǎlì

host country　东道国 dōngdàoguó

hot issue　热门股票 rèmén gǔpiào

hot money　游资 yōuzī

household consumption　家庭消费 jiātíng xiāofèi

housing loan　住宅贷款 zhùzhái dàikuǎn

human resources　人力资源 rénlì zīyuán

I

idle cash　闲置现金 xiánzhì xiànjīn

immediate payment　立即付款 lìjí fùkuǎn

imitation goods　冒牌货 màopáihuò

immovable estate　不动产 bùdòngchǎn

impact　影响 yǐngxiǎng；冲击 chōngjī

import declaration　进口报单 jìnkǒu bàodān

import duty　进口税 jìnkǒushuì

import licence = import permit
　　进口许可证 jìnkǒu xǔkězhèng

import quota [IQ]　进口限额 jìnkǒu xiàn'é

import surcharge　进口附加税 jìnkǒu fùjiāshuì

import surplus　入超 rùchāo

importer　进口商 jìnkǒushāng

in arrears　拖欠 tuōqiàn，未付的 wèifùde

in bulk　散装 sǎnzhuāng；大批 dàpī

in cash　用现金 yòng xiànjīn

in the red　赤字 chìzì，亏损 kuīsǔn

inactive market　不活跃的市场
　　　　　　bù huóyuède shìchǎng

inactive stock　呆滞存货 dāizhì cúnhuò；
　　　　　　冷门股票 lěngmén gǔpiào

incentive　奖励 jiǎnglì

inch　英寸 yīngcùn，吋 cùn

income 收入 shōurù，收益 shōuyì，所得 suǒdé

income tax [I/T] 所得税 suǒdéshuì

inconvertible currency
不能自由兑换的货币
bùnéng zìyóu duìhuàn de huòbì

increase 增加 zēngjiā，提高 dígāo

increment 增量 zēngliàng；增值 zēngzhí；
增加工资 zēngjiā gōngzī

indemnity 赔偿 péicháng，补偿金 bǔchángjīn

index 指数 zhǐshù；索引 suǒyǐn

indicator 指标 zhǐbiāo

indirect tax 间接税 jiànjiēshuì

individual income 个人所得 gèrén suǒdé

industrial dispute   劳资纠纷 láozī jiūfēn

industrial products   工业产品 gōngyè chǎnpǐn

inflation   通货膨胀 tōnghuò péngzhàng

information   资讯 zīxùn，信息 xìnxī，情报 qíngbào

initial capital   创办资本 chuàngbàn zīběn

inland revenue   国内税收 guónèi shuìshōu

inquiry   询问 xúnwèn；询价 xúnjià

instalment   分期付款 fēnqī fùkuǎn

instruction   指令 zhǐlìng，指示 zhǐshì；须知 xūzhī；说明书 shuōmíngshū

insurance   保险 bǎoxiǎn；
保险业务 bǎoxiǎn yèwù

insurance claim 保险索赔 bǎoxiǎn suǒpéi

insurance clause 保险条款 bǎoxiǎn tiáokuǎn

insurance company 保险公司 bǎoxiǎn gōngsī

insurance indemnity 保险赔偿 bǎoxiǎn péicháng

insurance policy [I/P] 保险单 bǎoxiǎndān,
保单 bǎodān

insurance premium 保险费 bǎoxiǎnfèi,
保费 bǎofèi

insured amount 保险金额 bǎoxiǎn jīn'é,
投保金额 tóubǎo jīn'é

inter-bank rate 银行同业利率
yínháng tóngyè lìlǜ

interest 利息 lìxī; 利益 lìyì; 股份 gǔfèn

interest-free credit 无息贷款 wúxī dàikuǎn

interest rate 利率 lìlǜ

interest rebate 利息回扣 lìxī huíkòu

interim report 期中报告 qīzhōng bàogào

internal trade 国内贸易 guónèi màoyì，
内贸 nèimào

international currency 国际货币 guójì huòbì

international direct dialing [IDD]
国际直拨长途电话
guójì zhíbō chángtú diànhuà

International Finance Corporation [IFC]
国际金融公司 Guójì Jīnróng Gōngsī

international financial market
国际金融市场 guójì jīnróng shìchǎng

International Monetary Fund [IMF]
国际货币基金组织
Guójì Huòbì Jījīn Zǔzhī

International Organization for Standardization
[IOS]　国际标准化组织
Guójì Biāozhǔnhuà Zǔzhī

international trade　国际贸易 guójì màoyì

Internet　网际网络 wǎngjì wǎngluò

interview　会见 huìjiàn，约见 yuējiàn；
面试 miànshì

introduction　介绍 jièshào；引进 yǐnjìn

inventory　存货 cúnhuò，盘存 páncún；
存货清单 cúnhuò qīngdān

investment　投资 tóuzī，投入资金 tóurù zījīn

investor　投资者 tóuzīzhě

invoice [inv.]　发票 fāpiào

inward charges　入港费 rùgǎngfèi

irrevocable letter of credit
　　　不可撤销的信用证
　　　bùkě chèxiāo de xìngyòngzhèng

issue　发行 fāxíng

issue price　发行价格 fāxíng jiàgé

J

jack up the price   抬高价格 táigāo jiàgé

job   工作 gōngzuò；职业 zhíyè；职位 zhíwèi

jobless   失业 shīyè

joint enterprise   联合企业 liánhé qǐyè，
合资企业 hézī qǐyè

joint signature   联合签署 liánhé qiānshǔ

joint venture   合资经营 hézī jīngyíng

juice   高利贷 gāolìdài；高利 gāolì

K

keen demand　　迫切需求 pòqiè xūqiú

kerb market　　场外交易市场
　　　　　　　chǎngwài jiāoyì shìchǎng

key industry　　基础工业 jīchǔ gōngyè

key personnel　　关键人员 guānjiàn rényuán

kickback　　回扣 huíkòu；茶钱 cháqián

knock-down price　　杀价 shājià

know-how　　技能 jìnéng；专门技术的知识
　　　　　　zhuānmén jìshù de zhīshi

L

label   标签 biāoqiān；标记 biāojì

labour cost   人工成本 réngōng chéngběn，
劳工成本 láogōng chéngběn

labour dispute   劳资争议 láozī zhēngyì

labour market   劳动力市场 láodònglì shìchǎng，
劳工市场 láogōng shìchǎng

land freight   陆运费 lùyùnfèi

land law   土地法 tǔdìfǎ

landing   卸货 xièhuò，起货上岸 qǐhuò shàng'àn

lapse   期满失效 qīmǎn shīxiào；权利终止 quánlì zhōngzhǐ

launch   开办 kāibàn；启用 qǐyòng

lawyer 律师 lǜshī

lay off 解雇 jiěgù

leading article 特价商品 tèjià shāngpǐn

leaflet 广告单 guǎngàodān,
宣传单 xuānchuándān

lease 租约 zūyuē，租赁 zūlìn

leasehold 租借权 zūjièquán

legacy duty 遗产税 yíchǎnshuì

legal advisor 法律顾问 fǎlǜ gùwèn

legal liability 法定债务 fǎdìng zhàiwù

legal proceedings 法律诉讼 fǎlǜ sùsòng；
诉讼程序 sùsòng chéngxù

legal right　法律权利 fǎlù quánlì

letter of advice　通知函 tōngzhīhán；
汇票通知单 huìpiào tōngzhīdān

letter of credit [L/C]　信用证 xìnyòngzhèng，
信用状 xìnyòngzhuàng

letter of guarantee [L/G]　保证书 bǎozhèngshū，
担保书 dānbǎoshū

letter of recommendation　推荐信 tuìjiànxìn，
介绍信 jièshàoxìn

levy　征税 zhēngshuì；征收 zhēngshōu

liability　负债 fùzhài；责任 zérèn

licence　许可 xǔkě；许可证 xǔkězhèng；
执照 zhízhào

lien　留置权 liúzhìquán，扣押权 kòuyāquán

life assurance = life insurance
人寿保险 rénshòu bǎoxiǎn

limit　限价 xiànjià；限制 xiànzhì

limited company　有限公司 yǒuxiàn gōngsī

line　种类 zhǒnglèi；行业 hángyè；
航线 hángxiàn

line of credit　信贷限额 xìndài xiàn'é

liner　班机 bānjī；定期客轮 dìngqī kèlún，
班轮 bānlún

liquid assets　流动资产 liúdòng zīchǎn

liquidate　清算 qīngsuàn；清偿 qīngcháng

list of exchange rate quotations
外汇牌价表 wàihuì páijiàbiǎo

listed company　上市公司 shàngshì gōngsī，
挂牌公司 guàpái gōngsī

listing　上市 shàngshì，挂牌 guàpái

litre　公升 gōngshēng

living allowance　生活津贴 shēnghuó jīntiē

living expense　生活费 shēnghuófèi

loading and unloading charges
装卸费 zhuāngxièfèi

loan　贷款 dàikuǎn，借款 jièkuǎn；
借出 jièchū

loan shark　高利贷者 gāolìdàizhě

loan without security　无担保贷款
wúdānbǎo dàikuǎn

local market   本地市场 běndì shìchǎng

local product   本地产品 běndì chǎnpǐn

local retailer   本地零售商 běndì língshòushāng

local tax   地方税 dìfāngshuì

local wholesaler   本地批发商 běndì pīfāshāng

location   出租 chūzū；位置 wèizhì

loco = on spot   当地交货价 dāngdì jiàohuòjià

lodge a claim   提出索赔 tíchū suǒpéi

long   多头 duōtóu；超买 chāomǎi

long-term loan   长期贷款 chángqī dàikuǎn

loose leaf   活页 huóyè

loss   亏损 kuīsǔn，损失 sǔnshī

lot   批 pī，宗 zōng

low-interest rate   低利率 dīlìlǜ

lump-sum payment   一次总付 yīcì zǒngfù

luncheon voucher   午餐券 wǔcānquàn

luxury tax   奢侈品税 shēchǐpǐnshuì

M

macro-economics 宏观经济学
　　　　　　　　　hóngguān jīngjìxué

made-up price 场外价格 chǎngwài jiàgé

mail bag 邮袋 yóudài

mail box 邮筒 yóutǒng，邮箱 yóuxiāng

mail order 邮购 yóugòu

mail transfer [M.T.] 信汇 xìnhuì

maintenance 维修 wéixiū，保养 bǎoyǎng；
　　　　　　　保证金 bǎozhèngjīn

make a deal 成交 chéngjiāo

make up 弥补 míbǔ；结算 jiésuàn

make-up   内部包装 nèibù bāozhuāng

maladjustment   失调 shītiáo,
调整不良 tiáozhěng bùliáng

maldistribution   分配不当 fēnpèi bùdàng

manage   经营 jīngyíng，管理 guǎnlǐ,
安排 ānpái；控制 kòngzhì

management board   管理委员会
guǎnlǐ wěiyuánhuì

management policy   管理决策 guǎnlǐ juécè,
经营方针
jīngyíng fāngzhēn

manager   经理 jīnglǐ

managing director   常务董事 chángwù dǒngshì

manipulation 操纵市场 cāozòng shìchǎng；
篡改 cuàngǎi

manufacture 制造 zhìzào；加工 jiāgōng；
制造业 zhìzàoyè

manufacturer 制造商 zhìzàoshāng，
厂家 chǎngjiā

margin 赚头 zhuàntóu，毛利 máolì；
边际 biānjì

mark-down 降价 jiàngjià

mark-up 提高标价 tígāo biāojià

market analysis 市场分析 shìchǎng fēnxī

market demand 市场需求 shìchǎng xūqiú

market fluctuation 市场波动 shìchǎng bōdòng

market price   市场价格 shìchǎng jiàgé

market value   市场价值 shìchǎng jiàzhí；
市价 shìjià，时价 shíjià

marketing   销售 xiāoshòu；市场学 shìchǎngxué

mass media   大众传播媒介
dàzhòng chuánbō méijiè

mass production   成批生产 chángpī shēngchǎn，
大规模生产
dàguīmó shēngchǎn

maturity   期满 qīmǎn；到期日 dàoqīrì

maximum [max.]   最高限度 zuìgāo xiàndù，
最高限价 zuìgāo xiànjià

measurement and weight list
容积重量表
róngjī zhòngliàng biǎo

media　宣传媒介 xuānchuán méijiè

membership　会员资格 huìyuán zīgé

memorandum [memo.]　备忘录 bèiwànglù；
便函 biànhán

merchandise　商品 shāngpǐn

merchant　商人 shāngrén

merger　合并 hébìng

metre　公尺 gōngchǐ，米 mǐ

metric system　公制 gōngzhì，十进制 shíjìnzhì

metric ton　公吨 gōngdūn

micro-economics　微观经济学 wēiguān jīngjìxué

middle rate　中间价 zhōngjiānjià,
平均价 píngjūnjià

miniboom　短暂繁荣 duǎnzàn fánróng

minimum charge　最低费用 zuìdī fèiyòng

minutes of meeting　会议记录 huìyì jìlù

miscellaneous expenses　杂费 záfèi,
杂项支出 záxiàng zhīchū

mobile shop　流动商店 liúdòng shāngdiàn

modernization　现代化 xiàndàihuà

monetary circulation　货币流通 huòbì liútōng

monetary control　金融管制 jīnróng guǎnzhì

monetary ease　银根松驰 yíngēn sōngchí

monetary fluctuation   货币波动 huòbì bōdòng

monetary unit   货币单位 huòbì dānwèi

money   货币 huòbì，钱 qián

money bloc   货币集团 huòbì jítuán

money changer   货币兑换商
huòbì duìhuànshāng；
钱币兑换机 qiánbì duìhuànjī

money market   货币市场 huòbì shìchǎng，
金融市场 jīnróng shìchǎng

money order [M.O.]   汇票 huìpiào，
汇款单 huìkuǎndān，
邮政汇票 yóuzhèng huìpiào

monopolize   垄断 lǒngduàn，专营 zhuānyíng

monopoly   垄断 lǒngduàn，专利 zhuānlì，
独家生意 dūjiā shēngyì

mortgage   抵押 dǐyā

most favoured nation [MFN]   最惠国 zuìhuìguó

motion   动议 dòngyì，方案 fāng'àn

movables   动产 dòngchǎn

multilateral trade   多边贸易 duōbiān màoyì

multinational company   跨国公司 kuàguó gōngsī

mutual interest   共同利益 gòngtóng lìyì

N

national income　国民收入 guómín shōurù

national treasury　国库 guókù

negotiable　可转让 kězhuǎnràng；
可谈判 kětánpàn

negotiation　谈判 tánpàn；议付 yìfù

net earnings　净收益 jìngshōuyì

net loss　净亏 jìngkuī，纯损 chúnsǔn

net price　实价 shíjià，净价 jìngjià

net profit　纯利润 chúnlìrùn，净利 jìnglì

net value　净值 jìngzhí

net weight [nt. wt.]　净重 jìngzhòng

new issue 新发行股 xīn fāxínggǔ

nominal rate of exchange 名义汇率
 míngyì huìlǜ, 挂牌汇率 guàpái huìlǜ

nominal value 票面价值 piàomiàn jiàzhí

non-dutiable goods 无税物品 wúshuì wùpǐn

non-member rate 非会员运费率
 fēihuìyuán yùnfèilǜ

non-monetary investment
 非货币性投资 fēihuòbìxìng tóuzī

non-profit organization
 非营利机构 fēiyínglì jīgòu

non-transferable 不可转让 bùkě zhuǎnràng

normal price 正常价格 zhèngcháng jiàgé

not for sale   非卖品 fēimàipǐn

note   票据 piàojù；纸币 zhǐbì；便条 biàntiáo

notice of dishonour   拒付通知 jùfù tōngzhī，
退票通知 tuìpiào tōngzhī

notification   通知书 tōngzhīshū；通知 tōngzhī

null and void   无效 wúxiào，作废 zuòfèi

numerical order   数目顺序 shùmù shùnxù

O

obligation 义务 yìwù；协议 xiéyì；
债务 zhàiwù

obsolete stocks 陈废存货 chénfèi cúnhuò，
过时存货 guòshí cúnhuò

odd lot 零星股票 língxīng gǔpiào，散股 sǎngǔ

offer 发盘 fāpán，发价 fājià，报价 bàojià

office hours 办公时间 bàngōng shíjiān；
营业时间 yíngyè shíjiān

office supplies 办公室用品 bàngōngshì yòngpǐn

official gold price 黄金官价 huángjīn guānjià

official price 官定价格 guāndìng jiàgé，
正式价格 zhèngshì jiàgé

official rate 官方汇价 guānfāng huìjià,
法定汇价 fǎdìng huìjià

official receipt 正式收据 zhèngshì shōujù

official record 正式记录 zhèngshì jìlù

official visa 公务签证 gōngwù qiānzhèng

offset 抵销 dǐxiāo

omission 遗漏 yílòu

on credit 赊帐 shēzhàng,
信用交易 xìnyòng jiāoyì

on probation 试用 shìyòng

on the job training 在职培训 zàizhí péixùn

one-man business 独资经营 dúzī jīngyíng,
独资生意 dúzī shēngyì

open an account　开立帐户 kāilì zhànghù，
开立户头 kāilì hùtóu

open bid　公开招标 gōngkāi zhāobiāo；
公开投标 gōngkāi tóubiāo

open market　自由市场 zìyóu shìchǎng，
公开市场 gōngkāi shìchǎng

open policy　开放政策 kāifàng zhèngcè；
预约保险单 yùyuē bǎoxiǎndān

open rate　自由运费率 zìyóu yùnfèilǜ

operation cost　营业成本 yíngyè chéngběn

operator　经营人 jīngyíngrén；
接线员 jiēxiànyuán

option　选择权 xuǎnzéquán；
购买或出售选择权
gòumǎi huò chūshòu xuǎnzéquán

oral agreement　口头协议 kǒutóu xiéyì

order　定货 dìnghuò，订购 dìnggòu；
　　　订货单 dìnghuòdān，
　　　订购单 dìnggòudān；
　　　汇款单 huìkuǎndān

order bill of lading [O.B/L]
　　指示提单 zhǐshì tídān

order cheque　记名支票 jìmíng zhīpiào

ordinary mail　平信 píngxìn，平邮 píngyóu

ordinary shares　普通股 pǔtōnggǔ

organization　组织 zǔzhī，机构 jīgòu

ounce [oz]　盎司 àngsī，英两 Yīngliǎng

out of stock [O/S]　无存货 wúcúnhuò，
　　　　　　缺货 quēhuò，脱销 tuōxiāo

output   产量 chǎnliàng；输出 shūchū

outstanding   未完成的 wèiwánchéngde；
未付款的 wèifùkuǎnde

outstanding balance   未付清余额 wèifùqīng yú'é

outstanding loan   未清贷款 wèiqīng dàikuǎn

outward port charges
出港手续费 chūgǎng shǒuxùfèi

overbuy   超买 chāomǎi

overdraft [O.D.]   透支 tòuzhī

overdue bill   过期票据 guòqī piàojù

over-estimate   估计过高 gūjì guògāo

over-valuation   估价过高 gūjià guògāo

overseas business　海外业务 hǎiwài yèwù

oversell　超卖 chāomài

oversubscribed　超额认购 chāo'é rèngòu

overtime [O.T.]　加班 jiābān

P

package deal　整套交易 zhěngtào jiāoyì，
成套买卖 chéngtào mǎimài

packaging　包装法 bāozhuāngfǎ；
包装设计 bāozhuāng shèjì

packing　包装 bāozhuāng，打包 dǎbāo

paid-up capital　实缴股本 shíjiǎo gǔběn

paper loss　帐面损失 zhàngmiàn sǔnshī

paper money　纸币 zhǐbì；票据 piàojù

paper profit　帐面利润 zhàngmiàn lìrùn

par value　票面价值 piàomiàn jiàzhí

parcel post [P.P.]　邮包 yóubāo；
包裹邮件 bāoguǒ yóujiàn

parent company　母公司 mǔgōngsī；
　　　　　　　　　　总公司 zǒnggōngsī

parity price　平价 píngjià；等价 děngjià

partial acceptance　部分承兑 bùfèn chéngduì

partial payment　部分支付 bùfèn zhīfù

partner　合伙人 héhuǒrén

partnership　合伙关系 héhuǒ guānxì，
　　　　　　　　　合股关系 hégǔ guānxì

passbook　银行存折 yínháng cúnzhé

passenger　乘客 chéngkè

passenger fare　客运费 kèyùnfèi

patent　专利 zhuānlì；专利权 zhuānlìquán

patent pool   专利共享 zhuānlì gòngxiǎng

pay by instalments   分期付款 fēnqī fùkuǎn

payment on delivery [P.O.D.]
       货到付款 huòdào fùkuǎn，
       交货付款 jiāohuò fùkuǎn

payment order   付款通知 fùkuǎn tōngzhī，
                  付款委托书 fùkuǎn wěituōshū

payroll   工资表 gōngzībiǎo

peak season   旺季 wàngjì

penalty   罚金 fájīn，罚款 fákuǎn

pension   退休金 tuìxiūjīn；养老金 yǎnglǎojīn

per annum [p.a.]   每年 měinián

per cent   百分之 bǎifēn zhī，巴仙 bāxiān

percentage   百分比 bǎifēnbǐ；百分率 bǎifēnlǜ

peril clause   危险条款 wēixiǎn tiáokuǎn

personal account [P/A]
个人帐户 gèrén zhànghù，
个人户头 gèrén hùtóu

personal loan   个人贷款 gèrén dàikuǎn

petty cash [P/C]   零用现金 língyòng xiànjīn

piece rate   计件工资 jìjiàn gōngzī

pledge   抵押品 dǐyāpǐn；抵押 dǐyā

port charges   港口费用 gǎngkǒu fèiyòng

port dues   入港费 rùgǎngfèi；入港税 rùgǎngshuì

port surcharge   港口附加税 gǎngkǒu fùjiāshuì

portfolio 有价证券 yǒujià zhèngquàn；
投资额 tóuzī'é

Port of Singapore Authority [PSA]
新加坡港务局 Xīnjiāpō Gǎngwùjú

post office 邮政局 yóuzhèngjú

post parcel [P/P] 邮寄包裹 yóujì bāoguǒ

postcard 明信片 míngxìnpiàn

postage 邮费 yóufèi

Postal Giro 邮政转帐 yóuzhèng zhuǎnzhàng

postal money order 邮政汇票 yóuzhèng huìpiào

postal remittance 邮政汇款 yóuzhèng huìkuǎn

poster 招贴 zhāotiē，海报 hǎibào

potential market　潜在市场 qiánzài shìchǎng

pound [lb]　磅 bàng

preference　优先权 yōuxiānquán；特惠 tèhuì

preference shares　优先股 yōuxiāngǔ

preferential tariff　优惠税则 yōuhuì shuìzé；
　　　　　　　　　特惠税率 tèhuì shuìlǜ

premium　升水 shēngshuǐ；奖金 jiǎngjīn；
　　　　　保险费 bǎoxiǎnfèi

president　董事长 dǒngshìzhǎng；总裁 zǒngcái

press release　发布新闻 fābù xīnwén

price list [P/L]　价表 jiàbiǎo，价目表 jiàmùbiǎo

pricing　订价 dìngjià；标价 biāojià

primary industry 基本工业 jīběn gōngyè

private placing 私下转销 sīxià zhuǎnxiāo

private property 私人财产 sīrén cáichǎn

pro rata 按比例 àn bǐlì

procurement 采购 cǎigòu

product 产品 chǎnpǐn

production cost 生产成本 shēngchǎn chéngběn

productivity 生产力 shēngchǎnlì；
生产率 shēngchǎnlǜ

professional etiquette 专业道德 zhuānyè dàodé

profit and loss account 损益帐 sǔnyìzhàng

profit taking　见利抛售 jiànlì pāoshòu，
套利 tàolì

profiteering　牟取暴利 móuqǔ bàolì

profits　利润 lìrùn；盈利 yínglì

progress report　进度报告 jìndù bàogào

promotion　推销 tuīxiāo，促销 cùxiāo；
擢升 zhuóshēng

property tax　财产税 cáichǎnshuì，
产业税 chǎnyèshuì

proprietor　业主 yèzhǔ

prospectus　说明书 shuōmíngshū；
发起书 fāqǐshū；
招股书 zhāogǔshū

protectionism　保护贸易主义
bǎohù màoyì zhǔyì

protocol　议定书 yìdìngshū

proxy　代理人 dàilǐrén；授权书 shōuquánshū；
委托书 wěituōshū

public auction　公开拍卖 gōngkāi pāimài

public property　公共财产 gōnggòng cáichǎn

purchase method　收购法 shōugòufǎ

purchase tax　购货税 gòuhuòshuì

pure (fine) gold　纯金 chúnjīn

pure (fine) silver　纯银 chúnyín

pyramid selling　金字塔推销法 jīnzìtǎ tuīxiāofǎ

Q

quality   品质 pǐnzhì，质量 zhìliàng

quality control [QC]   质量管理 zhìliàng guǎnlǐ，
    品质管制 pǐnzhì guǎnzhì

quantity [qt.]   数量 shùliàng

quarantine certificate   检疫证书 jiǎnyì zhèngshū

quasi-contract   准契约 zhǔngqìyuē

questionnaire   问卷 wènjuàn，问题单 wèntídān

quota system   限额制 xiàn'ézhì，配额制 pèi'ézhì

quotation   报价 bàojià；估价单 gūjiàdān

R

railway transportation   铁路运输 tiělù yùnshū

raising fund   集资 jízī，筹募资金 chóumù zījīn

range of prices   价格范围 jiàgé fànwéi，
价格幅度 jiàgé fúdù

rate of discount   贴现率 tiēxiànlǜ

rate of inflation   通货膨胀率
tōnghuò péngzhànglǜ

rate of premium   升水率 shēngshuǐlǜ；
保险费率 bǎoxiǎnfèilǜ

ratio   比率 bǐlǜ

raw material   原料 yuánliào

real estate 不动产 bùdòngchǎn，
房地产 fángdìchǎn

realization 变现 biànxiàn

ream [Rm.] 令 lǐng

rebate 回扣 huíkòu

receipt 收据 shōujù；收入 shōurù

receiver 收货人 shōuhuòrén；
收款人 shōukuǎnrén；
收件人 shōujiànrén

reciprocal trade agreement
互惠贸易协定
hùhuì màoyì xiédìng

recommend 推荐 tuíjiàn

recommendation 建议书 jiànyìshū

recovery 回收 huíshōu，追偿 zhuīcháng；
复苏 fùsū

recruitment 征聘职员 zhēngpìn zhíyuán

redemption 偿还 chánghuán，赎回 shúhuí

reference 证明人 zhèngmíngrén；案号 ànhào

refund 退款 tuìkuǎn；再筹资 zàichóuzī

register 登记 dēngjì，注册 zhùcè；
挂号 guàhào；登记簿 dēngjìbù

registered capital 注册资本 zhùcè zīběn

registered mail 挂号邮件 guàhào yóujiàn

registration fee 注册费 zhùcèfèi；
更名费 gēngmíngfèi

refrigerated goods 冷藏商品 lěngcáng shāngpǐn

regulations　规则 guīzé；法规 fǎguī

reimburse　偿还 chánghuán；偿付 chángfù

reinsurance [R.I.]　再保 zàibǎo

remedy　补救 bǔjiù

reminser　取佣经纪人 qǔyòng jīngjìrén

remittance　汇款 huìkuǎn

remitter　汇款人 huìkuǎnrén

renewal　续订 xùdìng；更新 gēngxīn；
展期 zhǎnqī

rental　租金 zūjīn；租金收入 zūjīn shōurù

reserves　储备金 chǔbèijīn

retail business　零售业务 língshòu yèwù

retail price 零售价格 língshòu jiàgé

retirement 赎回 shúhuí；退股 tuìgǔ；
退休 tuìxiū

retrenchment 裁员 cáiyuán

return 收益 shōuyì；（所得税）申报单
(suǒdéshuì) shēnbàodān

revaluation 升值 shēngzhí；重估 chónggū

revenue 岁入 suìrù，岁收 suìshōu

revenue tax 印花税 yìnhuāshuì

revocable letter of credit
可撤销信用证
kěchèxiāo xìnyòngzhèng

rights 权利 quánlì；认股权 rèngǔquán

risk of breakage   破损险 pòsǔnxiǎn

risk of leakage   渗漏险 shènlòuxiǎn

rollback   压价 yājià

round figures   整数 zhěngshù

route   航线 hángxiàn

royalty   版税 bǎnshuì；租费 zūfèi

S

safe deposit   保险库 bǎoxiǎnkù

sale by auction   拍卖 pāimài

sale by proxy   代销 dàixiāo

sales commission   销货佣金 xiāohuò yòngjīn

sales department   销售部 xiāoshòubù

sales force   推销人员 tuīxiāo rényuán

sales promotion   促销 cùxiāo,
推销活动 tuīxiāo huódōng

sales tax   销售税 xiāoshòushuī

sample   样品 yàngpǐn, 货样 hùoyàng;
样本 yàngběn

sample kit   成套样品 chéngtào yàngpǐn

savings account   储蓄帐户 chǔxù zhànghù

scrip   股票临时收据 gǔpiào línshí shōujù；
代价券 dàijiàquàn

scripless share   无票股份 wúpiào gǔfèn

sea mail   海上邮递 hǎishǎng yóudì

sea risk   海险 hǎixiǎn

secretariat   秘书处 mìshūchù

secular growth   持续增长 chíxù zēngzhǎng

secured loan   有担保贷款 yǒudānbǎo dàikuǎn，
抵押贷款 dǐyā dàikuǎn

securities   证券 zhèngquàn

securities exchange 证券交易所
zhèngquàn jiāoyìsuǒ

self-employed person 自营者 zìyíngzhě；
自雇人士 zìgù rénshì

self-service 自助 zìzhù，自助服务 zìzhù fúwù

sellers' market 卖方市场 màifāng shìchǎng

seller's option [s/o] 卖方选择 màifāng xuǎnzé

selling techniques 推销技术 tuīxiāo jìshù

semi-skilled worker 半熟练工人
bànshúliàn gōngrén

service industry 服务业 fúwùyè

settlement of accounts 结帐 jiézhàng

share 股份 gǔfèn

share broker　股票经纪人 gǔpiào jīngjìrén

share index　股票指数 gǔpiào zhǐshù

share option　股票购买权 gǔpiào gòumǎiquán

share premium　股票溢价 gǔpiào yìjià

shareholder　股票持有人 gǔpiào chíyǒurén，
股东 gǔdōng

shift duty　轮班职务 lúnbān zhíwù

shipment　装运 zhuāngyùn；船货 chuánhuò

shipping documents　装运单据 zhuāngyùn dānjù

shopping centre　购物中心 gòuwù zhōngxīn

shopping mall　购物商场 gòuwù shāngcháng

short　短少 duǎnshǎo；空头交易 kōngtóu jiāoyì

short covering　空头补进 kōngtóu bǔjìn

short sale　卖空交易 màikōng jiāoyì

short-term loan　短期贷款 duǎnqī dàikuǎn

showcase　陈列柜 chénglièguì

sight rate　即期汇率 jíqī huìlǜ

sign　签名 qiānmīng；签署 qiānshǔ；
标志 biāozhì

sign on　签约受雇 qiānyuē shòugù

signature　签字 qiānzì；印鉴 yìnjiàn

silver bar　银条 yíntiáo

silver coin　银币 yínbì

silver standard　银本位 yínběnwèi

simple interest   单利 dānlì

skilled worker   熟练工人 shúliàn gōngrén；
技工 jìgōng

slack season   淡季 dànjì

smuggled goods   走私货 zǒusīhuò

smuggling ring   走私集团 zǒusī jítuán

snap check   突击检查 tūjī jiǎnchá

snap decision   仓促的决策 cāngcùde juécè

social welfare   社会福利 shèhuì fúlì

software   软件 ruǎnjiàn

sold out   售完 shòuwán

sole agent   独家代理商 dújiá dàilǐshāng

source of supply　货源 huòyuán

solvency　偿付能力 chángfù nénglì

spare parts　零件 língjiàn

special account　特别帐户 tèbié zhànghù

special clause　特别条款 tèbié tiáokuǎn

specifications　规格 guīgé；货品说明书 huòpǐn shuōmíngshū

speculation　投机 tóujī

split-up　拆股 chāigǔ

sponsor　赞助者 zànzhùzhě；主办人 zhǔbànrén

spot price　现货价格 xiànhuò jiàgé

spot transaction　现货交易 xiànhuò jiāoyì

squeeze　　挤兑 jǐduì；紧缩银根 jǐnsuō yíngēn

squeeze the shorts　　杀空头 shā kōngtóu

staff　　职员 zhíyuán，工作人员 gōngzuò rényuán

stamp　　图章 túzhāng；印花 yìnhuā；
邮票 yóupiào

stamp duty　　印花税 yìnhuāshuì

standard of living　　生活水平 shēnghuó shuǐpíng

standardization　　标准化 biāozhǔnhuà

standby　　备用品 bèiyòngpǐn；
备用设备 bèiyòng shèbèi

standing cost　　经常费用 jīngcháng fèiyòng；
长期成本 chángqī chéngběn

state enterprise　　国营企业 guóyíng qǐyè

statement of account 帐户结算单
zhànghù jiésuàndān

statement of assets and liabilities
资产负债表 zīchǎn fùzhàibiǎo

statistical data 统计数据 tǒngjì shùjù，
统计资料 tǒngjì zīliào

statutory rights 法定权利 fǎdìng quánlì

stevedore 码头装卸工人
mǎtou zhuāngxiè gōngrén；
搬运工人 bānyùn gōngrén

stock collateral loan 股票抵押贷款
gǔpiào dǐyā dàikuǎn，
证券担保贷款 zhèngquàn dānbǎo dàikuǎn

stock exchange 股票交易所 gǔpiào jiāoyìsuǒ，
证券交易所
zhèngquàn jiāoyìsuǒ

stock market 股票市场 gǔpiào shìchǎng,
证券市场 zhèngquàn shìchǎng

stock premium 股票升水 gǔpiào shēngshuǐ,
股票溢价 gǔpiào yìjià

stock price index 股票价格指数
gǔpiào jiàgé zhǐshù

stock taking 清货 qīnghuò,
清点存货 qīngdiǎn cúnhuò

strike a bargain 成交 chéngjiáo

strictly confidential 高度机密 gāodù jīmì,
绝密 juémì

subject to approval 须经批准 xūjīng pīzhǔn

subscribe 认购 rènggòu；捐助 juānzhù；
订阅 dìngyuè

subscription of shares　认购股票 rèngòu gǔpiào

subsidiary　子公司 zǐgōngsī；
附属机构 fùshǔ jīgòu；
补充的 bǔchōngde

successful bidder　得标人 débiāorén，
中标人 zhòngbiāorén

summary　摘要 zhāiyào，概要 gàiyào；
一览表 yīlǎnbiǎo

superior　上级 shàngjí，上司 shàngsī；
优越的 yōuyuède

supervisor　监督人 jiāndūrén，
主管人 zhǔguǎnrén

supplier　供应商 gōngyìngshāng

supply and demand　供求 gōngqiú；
供求关系 gōngqiú guānxì

surcharge　附加税 fùjiāshuì；附加费 fùjiāfèi

surety　保证 bǎozhèng；保证人 bǎozhèngrén

surplus　剩余 shèngyú，盈余 yíngyú；
　　　　　顺差 shūnchā

syndicate　银团 yíntuán，财团 cáituán

T

take-off period　经济起飞时期 jīngjì qǐfēi shíqī

takeover　兼并 jiānbìng，收购 shōugòu

tally charges　点数费 diǎnshǔfèi，
　　　　　　　点货费 diǎnhuòfèi

tangible assets　有形资产 yǒuxíng zīchǎn

tanker　油轮 yóulún

tare　皮重 pízhòng

target market　目标市场 mùbiāo shìchǎng

tariff　关税 guānshuì；收费表 shōufèibiǎo

tax evasion　偷税 tōushuì，漏税 lòushuì，
　　　　　　　逃税 táoshuì

tax-exemption　免税 miǎnshuì；
　　　　　　　免税额 miǎnshuì'é

tax payer　纳税人 nàshuìrén

technical decline　技术性下降 jìshùxìng xiàjiàng

technical rally　技术性回升 jìshùxìng huíshēng

telefacsimile　电话传真 diànhuà chuánzhēn

telegraphic transfer (T/T)　电汇 diànhuì

telex (TLX)　电传 diànchuán

temporary receipt　临时收据 línshí shōujù

tender　招标 zhāobiāo；投标 tóubiāo

term loan　定期贷款 dìngqī dàikuǎn

terminal 终点站 zhōngdiǎnzhàn；
转换站 zhuǎnhuànzhàn

testimonial 证明书 zhèngmíngshū；
介绍信 jièshàoxìn

third party 第三方 dìsānfāng

title deeds 所有权凭证 suǒyǒuquán píngzhèng

toll 通行费 tōngxíngfèi，通行税 tōngxíngshuì

toll-free line 免费电话线 miǎnfèi diànhuàxiàn

top up 填补 tiánbǔ

total amount 总数 zǒngshù

total cost 总成本 zǒngchéngběn

total loss (T.L.)　全损 quánsǔn

trade agreement　贸易协定 màoyì xiédìng

trade barrier　贸易壁垒 màoyì bìlěi

trade bloc　贸易集团 màoyì jítuán

trade deficit　贸易赤字 màoyì chìzì

trade fair　商品交易会 shāngpǐn jiāoyìhuì；
商品展览会 shāngpǐn zhǎnlǎnhuì

trade mark　商标 shāngbiāo

trade protectionism　贸易保护主义
màoyì bǎohù zhǔyì

trade sanction　贸易制裁 màoyì zhìcái

trade surplus　贸易顺差 màoyì shùnchā，
出超 chūchāo

transfer 转让 zhuǎnràng，过户 guòhù

transfer of technology 技术转让 jìshù zhuǎnràng

transfer of title 所有权转让
suǒyǒuquán zhuǎnràng

transferable 可转让 kězhuǎnràng

transit 过境运输 guòjìng yùnshū，
转口 zhuǎnkǒu；转机 zhuǎnjī

transit goods 过境货物 guòjìng huòwù，
转口货物 zhuǎnkǒu huòwù

transit visa 过境签证 guòjìng qiānzhèng

travel agency 旅行社 lǚxíngshè

travel documents 旅行证件 lǚxíng zhèngjiàn

treasury 国库 guókù

treaty　条约 tiáoyuē

trust　信托 xìntuō；托拉斯 tuōlāsī

trust fund　信托基金 xìntuō jījīn

trustee　受托人 shòutuōrén

turnover　营业额 yíngyè'é；成交额 chéngjiāo'é

U

ultimo[ult.]　上月的 shàngyuède

under licence　领有执照 lǐngyǒu zhízhào

undertake　承担 chéngdān；同意 tóngyì；
承办 chéngbàn；保证 bǎozhèng

undervaluation duty　低报税 dībàoshuì

underwrite　承保 chéngbǎo；认购 rèngòu；
签署 qiānshǔ

unearned income　非就业收入 fēijiùyè shōurù

unemployment　失业 shīyè

unfavourable balance　逆差 nìchā

uniform price　统一价格 tǒngyì jiàgé，
不二价 bù'èrjià

unit cost　单位成本 dānwèi chéngběn

unit price　单价 dānjià

unit trust　单位投资信托 dānwèi tóuzī xìntuō；
信托基金 xìntuō jījīn

unloading　卸货 xièhuò；倾销 qīngxiāo；
抛售 pāoshòu

unstamped　未贴印花 wèitiē yìnhuā

unused fund　未用资金 wèiyòng zījīn

upgrade　提升 tíshēng；栽培 zāipéi；
提高 tígāo

upswing　好转期 hǎozhuǎnqī

upward trend　涨势 zhǎngshì

utility　公用事业 gōngyòng shìyè

V

vacancy   空缺 kòngquē；空房间 kòngfángjiān

valid period   有效期 yǒuxiàoqī

validity   有效 yǒuxiào；合法性 héfǎxìng

valuable goods   贵重货物 guìzhòng huòwù

valuation   估价 gūjià；估值 gūzhí

valued customer   好客户 hǎo kèhù

venture capital   风险资本 fēngxiǎn zīběn；
创业资本 chuàngyè zīběn

verbal agreement   口头协定 kǒutóu xiédìng

verification of account   帐目核对 zhàngmù héduì

verify   检验 jiǎnyàn；核对 héduì

Visa card　信用卡 xìnyòngkǎ

volume of business　营业额 yíngyè'é；
成交量 chéngjiāoliàng

voucher　凭证 píngzhèng，传票 chuánpiào，
凭单 píngdān

W

wage level　工资水平 gōngzī shuǐpíng

wage freeze　工资冻结 gōngzī dòngjié

war risk　战争险 zhànzhēngxiǎn

warehouse　仓库 cāngkù

warrant　认股证书 rèngǔ zhéngshū，
凭单 píngdān

warranty　保用单 bàoyòngdān；
保证书 bǎozhèngshū

weights and measures　度量衡 dùliànghéng

welfare　福利 fúlì

wharfage　码头费 mǎtoufèi

when issued [w.i.]　假若发行 jiǎruò fāxíng

white-collar workers　白领阶层 báilǐng jiēcéng

wholesale　批发 pīfā

wholesale dealer　批发商 pīfāshāng

wholesale price　批发价格 pīfā jiàgé

with effect from [w.e.f.]
　即日起生效 jírì qǐ shēngxiào

with particular average [W.P.A.]
　水渍险 shuǐzìxiǎn

withdrawal　提款 tíkuǎn；退股 tuìgǔ；
　撤回 chèhuí

without prejudice [W.P.]
　不损害权利 bù sǔnhài quánlì

work load   工作量 gōngzuòliàng,
工作负荷 gōngzuò fùhè

working capital   周转资本 zhōuzhuǎn zīběn

working hours   工作时间 gōngzuò shíjiān

World Trade Organization [WTO]
世界贸易组织 Shìjiè Màoyì Zǔzhī

write off   注销 zhùxiāo

Y

yield 收益 shōuyì；收益率 shōuyìlǜ

Z

zero growth　零增长 líng zēngzhǎng

zone　区域 qūyù；区 qū

附录
APPENDIX

Currencies   货币名称
huòbì míngchēng

Dollar   元   Yuán

Australian Dollar
澳大利亚元
Àodàlìyà Yuán

Canadian Dollar
加拿大元
Jiānádà Yuán

Ethopian Dollar
埃塞俄比亚元
Āisài'ébǐyā Yuán

Hong Kong Dollar
港元 Gǎng Yuán

Jamaican Dollar
牙买加元
Yámǎijiā Yuán

Liberian Dollar
利比里亚元
Lìbǐlǐyà Yuán

| New Zealand Dollar | 新西兰 (纽西兰) 元 |
| | Xīnxīlán (Niǔxīlán) Yuán |

| Singapore Dollar | 新加坡元 |
| | Xīnjiāpō Yuán |

| Trinidad and Tobayo Dollar | 特立尼达和多巴哥元 |
| | Tèlìnídá hé Duōbāgē Yuán |

| U.S. Dollar | 美元 Měi Yuán |

| Chinese Reminbi [Yuan] | 中国人民币〔元〕 |
| | Zhōngguó Rénmínbì [Yuán] |

| Japanese Yen | 日本圆 Rìběn Yuán |

| Malaysian Ringgit [Dollar] | 马来西亚零吉〔元〕 |
| | Mǎláixīyà Língjí [Yuán] |

Franc 法郎 Fǎláng

Belgian Franc
比利时法郎
Bǐlìshí Fǎláng

Burundi Franc
布隆迪法郎
Bùlóngdí Fǎláng

French Franc
法国法郎
Fǎguó Fǎláng

Luxembourg Franc
卢森堡法郎
Lúsēnbǎo Fǎláng

Malagasy Franc
马尔加什法郎
Mǎěrjiāshí Fǎláng

Rwanda Franc
卢旺达法郎
Lúwàngdá Fǎláng

Swiss Franc
瑞士法郎
Ruìshì Fǎláng

African Financial Community Franc (CFA Franc)	非洲金融共同体法郎 Fēizhōu Jīnróng Gòngtóngtǐ Fǎláng
Cameroon	喀麦隆 Kèmàilóng
Central African Republic	中非共和国 Zhōngfēi Gònghéguó
Chad	乍得 Zhàdé
Congo	刚果 Gāngguǒ
Dahomey	达荷美 Dáhéměi
Gabon	加蓬 Jiāpéng
Gambian Dalasi	冈比亚达拉西 Gāngbǐyà Dálāxī
Ghanaian Cedi	加纳塞地 Jiānà Sàidì

Guinean Syli	几内亚西里 Jīnèiyà Xīlǐ
Ivory Coast	象牙海岸 Xiàngyá Hǎi'àn
Mali	马里 Mǎlǐ
Niger	尼日尔 Nírìěr
Senegal	塞内加尔 Sàinèijiāěr
Togo	多哥 Duōgē
Upper Volta	上沃尔特 Shàng Wòěrtè

Mark 马克 mǎkè

Deutsche Mark, 德国马克
  Deutschemark Déguó Mǎkè

Finnish Markka 芬兰马克
Fēnlán Mǎkè

Pound 镑 Bàng

Sterling Pound 英镑
YīngBàng

Cyprus Pound 塞浦路斯镑
Saìpǔlùsī Bàng

Egyptian Pound 埃及镑
Āijí Bàng

Lebanese Pound 黎巴嫩镑
Líbānèn Bàng

Libyan Pound 利比亚镑
Lìbǐyà Bàng

Maltese Pound	马耳他镑 Mǎěrtā Bàng
Sudanese Pound	苏丹镑 Sūdān Bàng
Syrian Pound	叙利亚镑 Xùlìyà Bàng

Colon   科郎   Kēláng

Costa Rican Colon	哥斯达黎加科郎 Gēsīdálíjiā Kēláng
Salvadoran Colon	萨尔瓦多科郎 Sāěrwǎduō Kēláng

Dinar   第纳尔   Dìnàěr

Dinar of the People's Democratic Republic of Yemen	也门民主人民 共和国第纳尔 Yěmén Mínzhǔ Rénmín Gònghéguó Dìnàěr

Algerian Dinar	阿尔及利亚第纳尔 Āěrjílìyà Dìnàěr
Iraqi Dinar	伊拉克第纳尔 Yīlākè Dìnàěr
Jordanian Dinar	约旦第纳尔 Yuēdàn Dìnàěr
Kuwaiti Dinar	科威特第纳尔 Kēwēitè Dìnàěr
Tunisian Dinar	突尼斯第纳尔 Tūnísī Dìnàěr

Escudo 埃斯库多 Āisīkùduō

| Chilean Escudo | 智利埃斯库多
Zhìlì Āisīkùduō |
| Portuguese Escudo | 葡萄牙埃斯库多
Pútáoyá Āisīkùduō |

Krone　　克朗　Kèláng

Danish Krone
丹麦克朗
Dānmài Kèláng

Icelandic Krone
冰岛克朗
Bīngdǎo Kèláng

Norwegian Krone
挪威克朗
Nuówēi Kèláng

Swedish Krone
瑞典克朗
Ruìdiǎn Kèláng

Kwacha　　克瓦查　Kèwǎchǎ

Malawi Kwacha
马拉维克瓦查
Mǎlāwéi Kèwǎchǎ

Zambian Kwacha
赞比亚克瓦查
Zànbǐyà Kèwǎchǎ

Peso	**比索** Bǐsuǒ
Argentine Peso	阿根廷比索 Āgēntíng Bǐsuǒ
Bolivian Peso	玻利维亚比索 Bōlìwéiyà Bǐsuǒ
Colombian Peso	哥伦比亚比索 Gēlúnbǐyà Bǐsuǒ
Cuban Peso	古巴比索 Gǔbā Bǐsuǒ
Dominican Peso	多米尼加比索 Duōmǐníjiā Bǐsuǒ
Mexican Peso	墨西哥比索 Mòxīgē Bǐsuǒ
Philippine Peso	菲律宾比索 Fēilǜbīn Bǐsuǒ
Uruguayan Peso	乌拉圭比索 Wūlāguī Bǐsuǒ

Shilling	先令　Xiānlìng
Kenya Shilling	肯尼亚先令 Kěnníyà Xiānlìng
Somali Shilling	索马里先令 Suǒmǎlǐ Xiānlìng
Tanzania Shilling	坦桑尼亚先令 Tānsāngníyà Xiānlìng
Uganda Shilling	乌干达先令 Wūgāndá Xiānlìng
Rial	里亚尔　Lǐyàěr
Iranian Rial	伊朗里亚尔 Yīlǎng Lǐyàěr
Saudi Arabian Rial	沙特阿拉伯里亚尔 Shātè Ālābó Lǐyàěr
Yemeni Rial	阿拉伯也门里亚尔 Ālābó Yěmén Lǐyàěr

Rupee 卢比 Lúbǐ

Indian Rupee 印度卢比
Yìndù Lúbǐ

Maldives Rupee 马尔代夫卢比
Mǎěrdàifū Lúbǐ

Nepalese Rupee 尼泊尔卢比
Níbóěr Lúbǐ

Pakistan Rupee 巴基斯坦卢比
Bājīsītǎn Lúbǐ

Sri Lankan Rupee 斯里兰卡卢比
Sīlǐlánkǎ Lúbǐ

Others 其他 qítā

Afghani 阿富汗尼
Āfùhàn Ní

Albanian Lek 阿尔巴尼亚列克
Ā'ěrbāníyà Lièkè

Austrian Schilling	奥地利先令 Àodìlì Xiānlìng
Brazilian Cruzeiro	巴西克鲁赛罗 Bāxī Kèlǔsàiluó
Bulgarian Lev	保加利亚列弗 Bǎojiālìyà Lièfú
Cambodian Riel	柬埔寨瑞尔 Jiǎnbǔzhài Ruì'ěr
Czechoslovak Koruna	捷克斯洛伐克克朗 Jiékèsīluòfákè Kèláng
Ecuadoran Sucre	厄瓜多尔苏克雷 Èguāduōěr Sūkèléi
Greek Drachma	希腊德拉克马 Xīlà Délàkèmǎ
Guatemalan Quetzal	危地马拉格查尔 Wēidìmǎlā Géchá'ěr
Haitian Gourde	海地古德 Hǎidì Gǔdé

Honduran Lempira	洪都拉斯伦皮拉 Hóngdūlāsī Lúnpílā
Hungarian Forint	匈牙利福林 Xiōngyálì Fúlín
Indonesian Rupiah	印度尼西亚盾 Yìndùníxīyà Dùn
Italian Lira	意大利里拉 Yìdàlì Lǐlā
Korean Won	朝鲜圆 Cháoxiān Yuán
Laotian Kip	寮国基普 Liáoguó Jīpǔ
Macao Pataca	澳门元 Àomén Yuán
Mauritania Ougiya	毛里塔尼亚乌吉亚 Máolǐtāníyà Wūjíyà
Myanmar Kyat	缅甸元 Miǎndiàn Yuán